PAIDEIA MONOGRAPHS

THE ANALOGICAL
CONCEPTS

HERMAN
DOOYEWEERD

PAIDEIA
PRESS

www.paideiapress.ca

www.reformationaldl.org

The Analogical Concepts

Translation of *De analogische grondbegrippen der vakwetenschappen en hun betrekking tot de structuur van den menselijken ervarings-horizon. Mededelingen der Koninklijke Nederlandse Akademie van Wetenschappen, afd. Letterkunde,* New series, Volume 17, No. 6 (Amsterdam: Noord-Hollandsche Uitgevers Maatschappij, 1954).

This English edition, translated by Robert D. Knudsen, is a publication of Paideia Press (3248 Twenty First St., Jordan Station, Ontario, Canada L0R 1S0). Copyright ©2022 by Paideia Press. All rights reserved.

Except for brief quotations in critical publications or reviews, no part of this book may be reproduced in any manner without prior written permission from Paideia Press at the address above.

Unless otherwise indicated, Scripture quotations are from the ESV® Bible (The Holy Bible, English Standard Version®). Copyright © 2001 by Crossway, a publishing ministry of Good News Publishers. Used by permission. All rights reserved.

Paideia Monograph Series Editor: Steven R. Martins

Book Design by: Michael Wagner

ISBN 978-0-88815-319-7

Printed in the United States of America

Contents

The Interrelationship of Elementary
Basic Concepts   5

Other Examples of Analogical Concepts   15

Philosophical Neglect   34

About the Author   51

THE ANALOGICAL
CONCEPTS

The Interrelationship of Elementary Basic Concepts

THE SUBJECT BEFORE us concerns a problem that as yet has been largely ignored but which involves nevertheless the foundations of all the special sciences without exception. I refer to the nature and the mutual interrelationship of the elementary basic concepts which the various special sciences employ without giving an account of their peculiar meaning and mutual connections. At first sight these fundamental concepts do not appear to be bound to the particular aspects of human experience in terms of which the fields of investigation of the various special sciences are in principle distinguished from each other. They appear to transcend those boundaries and in a clear fashion to give expression to the unity of science above all the variety of points of view under which the particular sciences view empirical reality. Closer scrutiny teaches us, however, that in the various branches of science there is no such thing as an unambiguous employment of these basic concepts.

On the contrary, these concepts display an ambiguous or analogical character that cannot be laid at the door of an unfortunate use of words. And, if they are not more closely defined as to their meaning within a particular field of investigation, they can lead to serious impasses in the delineation of problems in the special sciences and in philosophy. The analogical concepts can obtain this precision of meaning only in terms of the general nature of the particular field of scientific investigation in which they are employed. Since these fields of investigation are characterized in principle by the various aspects of the horizon of human experience,[1] it is therefore these aspects themselves which must guarantee to the analogical concepts their

---

1. This is the case only with reference to real *special* sciences, not to sciences such as sociology and anthropology, whose actual field of investigation are various structures of individuality which arch over the various aspects of experience and bind them together into a typical whole. We cannot enter further into this matter in this connection.

THE ANALOGICAL CONCEPTS   7

particular *modal* sense. For the aspects in question are themselves nothing more than the fundamental ways or *modi* in terms of which we experience temporal reality. They may not be confused therefore with concrete phenomena; instead, they form a modal framework in which we grasp concrete phenomena from different points of view. These modes can be tracked down only by means of the theoretical analysis of the structure of our integral experience, in connection with which it is precisely the exact delineation of the meaning of the analogical concepts that renders important services. The modes are the following: quantity (number), spatiality, movement, energetic effect, organic life, feeling, logical distinction, historical or cultural development, symbolical meaning, social commerce, economic valuation, and in addition the aesthetic, juridical, and moral ways of experiencing, together with that of faith.

As soon as we attempt, however, to subject the modal sense of these aspects to a theoretical analysis, they manifest a strongly analogical structure.

That is to say, every aspect of experience expresses within its modal structure the entire temporal order and connection of all the aspects. Only the central moment (*kern-moment*) of its modal structure, what we may call the modal nucleus (*zinkern*) of the aspect, manifests here an original and univocal character. But it can express this irreducible nucleus of meaning of the aspect only in connection with a series of analogical moments of meaning, which, on the one hand, refer back to the nuclei of meaning of all the earlier aspects and, on the other hand, point forward to the nuclei of meaning of all the later ones. It is to these analogical moments in the modal structure of the various aspects of our experience that the analogical concepts of the various special sciences are related. They express therefore an inner interrelatedness between the various fields of science, but they cannot do away with their modal diversity of meaning. They must receive their modal qualification from the irreducible nucleus of meaning of that aspect of experience which establishes the general characteristic of the particular scientific field in question.

This state of affairs is expressed intuitively in the very terminology of the special sciences, in that to avoid misunderstanding one prefaces the terms referring to analogical concepts with an adjective, which indicates the general modal nature of the particular area of investigation.

Thus in physics one speaks of *physical* space, in biology of *life*-space (ecology) or *life*-milieu (*Umwelt*), in psychology of the space of *sensory perception*, in logic of *logical* extension or *formal-analytical* space, in jurisprudence of *juridical* space or the domain in which legal norms are valid, in economics of *economic* space, etc.

All of these analogical concepts of space are in the last analysis related to the nucleus of meaning of the spatial aspect: *extension*. Nevertheless *in the analogical use* of the latter there is something else than what is meant by pure spatiality in the original sense of unbroken dimensional extension in the complete simultaneity of all its points. Irregardless of whether this original spatiality is thought of metrically in a Euclidean or non-Euclidean way, it is not qualified as such in a physical, or biological, or sensory, or logical, or historical, or economical, or juridical fashion. This is what Newton had in mind in his idea of absolute space and what the Greeks already perceived when they distinguished the exact spatial relationships as pure form from all material and sensory perceptible things.

Undoubtedly Kant had the intuition of space in this original sense in mind when he declared that space was an *a priori* form of sensory experience. But while Newton gave absolute space a metaphysical interpretation, conceiving it as the *sensorium Dei*, for which reason he of course could recognize no other concepts of space, Kant, who just as much allowed for only one sense of space, ascribed to pure space a transcendental-psychological sense, allowing sensory impressions to appear in the pure form of a Euclidean space.

This is in fact impossible. The space of sensory experience—of seeing, touching, and hearing—is in principle different from pure mathematical space. In the pure form of the latter we can never receive sensory impressions of space. Hume's criticism of pure Euclidean geometry in its nonformalized sense is from a psychological point of view irrefutable.

Likewise it is impossible to reduce physical space to that of pure mathematics. Its characteristics are completely dependent upon energy, and since the quantum theory has shown that the radiation of energy is not continuous but is bound to quanta, legitimate doubt has arisen as to the continuity of physical space. Gravitational fields and electro-magnetic fields are not simply spatial entities and are not perceptible to the senses.

That also logical thought-space or formal-analytical space, as they are presented in *Logistik*[2], cannot be space in its pure original sense, is evident. This space is analytical in character; it is an order of logical

---

2. See Rudolf Carnap, *Der Raum, Ein Beitrag zur Wissenschaftslehre* (Berlin, 1922).

THE ANALOGICAL CONCEPTS 11

coexistence in which we localize every logical element. The logical extension of concepts and judgments is a truly *logical* extendedness; but this logical extendedness cannot be identical with the *original* intuition of spatiality, although it is related to it in an unbreakable coherence of meaning. Wherever the sense of the concept of space is dependent upon a non-spatial modal qualification, this concept does not have the character of the original concept of space and we have to do only with an *analogy* of space.

In historiography we work with an historical concept of space, the concept of a cultural area which in historical development manifests a dynamic and not a static character. The culture-historical aspect of our experiencing is qualified by the nucleus of meaning of the cultural modality, namely, a controlling fashion of giving form to social relationships according to a free project, which implies power over persons and over things. The concept of historical space has to do therefore with the extension of power, and the extension of power is not extension in the original sense of space, although it presupposes the latter.

A culture area is not identical either with a part of the field of sensory experience, even though in our experiencing it is unbreakably connected with it. Spheres of historical power as such cannot be experienced by the senses and they cannot function as such in the space of sensory experience any more than the *historical extendedness* of a sphere of power can. The idea that this can indeed be the case arises out of the confusion of the modal historical aspect of our experiencing with history in the concrete sense of what has actually occurred in the past. An event in the past functions in all of the aspects of our experiencing without exception in their unbreakable interconnectedness. Just as it is impossible for any other science to investigate a phenomenon in all of its concreteness, it is impossible for historiography to investigate history as to all of its modal aspects and teach us "how it really happened." It has as its special focus of interest the development of power relationships, and this in itself is a theoretical abstraction.

The battle of Waterloo is a unity if we consider it as an historically qualified fact, that is to say, if we consider it in terms of historical power as a decisive test of military strength between Napoleon and the allied powers which opposed him. Should we try to give an account, however, of everything that really took place there, we should also have to include the moral, juridical, economical, lingual, and social aspects of the event, the organic life-processes and the physical-chemical processes in the bodies of the fighting soldiers, the emotions, impressions, and thought associations that took place in each one of them during the battle, the changes in the atmosphere, the trajectories of the bullets, the reactions in the animal world, etc. In its concrete reality the battle of Waterloo, of course, also had its aspect of sensory experience; but looking at it exclusively from this point of view we should not be able to say what belonged and what did not belong to this historical event.

The well-known economist F. H. Hayek has asked the question whether the frantic efforts of the farmers to harvest the crops that were in the way of the approaching armies also belonged to the battle of Waterloo.[3]

This question is very instructive if we bring it into connection with the problem with which we are concerned, i.e., the concept of historical space. Within the space of sensory experience taken by itself the bounds of the battle of Waterloo as an historical fact cannot be established, because within this space it does not display any historical unity.

An historical field of battle depends upon a free project and is qualified by power relationships which as such have a super-sensory character.

---

3. F. H. Hayek, "The Facts of the Social Sciences," *Ethics*, LIV (1943), 1-13.

In our experience it is unbreakably connected with the space of sensory perception, but precisely in its modal historical sense it is not identical with it. That is no less the case with a language area, in which we are confronted with the linguistic concept of space. Here also we must distinguish sharply between the veritable linguistic aspect of our experiencing, that of symbolical signification related to the understanding of this signification by others, and a concrete act, for example, speech, making a deictic or mimic movement, or giving a signal. In principle the concrete act functions in all aspects of our experiencing. We can localize sounds, particular sensory figures or impressions within the space of sensory perception. But we cannot experience in a sensory fashion that they are symbols with a particular symbolic meaning, because signification is a super-sensory relationship which is different in principle from a sensory association. How would it be possible to localize languages in the space of sensory perception? The true state of affairs is this, that a language area is as such a super-sensory social quantity, an extensive sphere of mutual possibilities of understanding, a linguistic extension which indeed is unbreakably connected with the space of sensory perception (and beyond it with pure space) but which cannot be localized within space itself. A geographical delineation of a language area is itself not anything more than a symbolical designation of it.

The same is true of an area of juridical competency, for instance, the territory of a particular state. I can experience a piece of ground in a sensory fashion, but not a particular geographical area as a sphere of competency of a particular governmental authority. Such an area can only be signified in a symbolical fashion, because juridical life rests of necessity on a symbolical foundation but is not experienceable as such by the senses.

Who could experience in a purely sensory fashion that a ship sailing in the Atlantic Ocean under a particular national flag, or the embassy of a particular nation in London is territory of that particular nation? But the symbols with which this juridical analogy of space is signified must have as concrete things an aspect of sensory experience, because the linguistic aspect of our experiencing hangs together unbreakably with the sensory.

Other Examples of Analogical Concepts

I now turn to still other examples of analogical concepts. In the first place the concept of movement. Aristotle already discerned the analogical character of this concept, without relating it truly to the modal aspects of experience. According to him motion can mean change of place, qualitative change, and substantial change.

But analogies of movement must in the final analysis have to do with movement in a non-analogical sense. As analogies they are not recognizable apart from their ultimate connectedness with the original or pure aspect of movement of our experiencing.

Only pure dynamics as it was developed by Galileo can give us the concept of motion with its inherent principle of inertia in its original, non-analogical sense. This concept of motion cannot be dependent on the concept of power or energy. Pure motion which is not qualified by the nucleus of meaning of another aspect of our experiencing, cannot be defined as change of place in pure space. For this movement is a continuous, extensive flow in the succession of its moments; and in pure space it is not possible. In none of its moments is movement able to be localized in a pure mathematical space without dissolving it in a series of static points, that is to say, eliminating it. Change of place presupposes that movement *had* a place in a particular moment, what is precisely not the case. Nevertheless no movement as such is possible apart from an inner relationship to pure spatiality and this relationship expresses itself in the original

THE ANALOGICAL CONCEPTS 17

aspect of movement of our experiencing in a spatial analogy, the successive extension of the direction of movement which is unbreakably connected with the time of movement. Mathematically we can grasp this movement only in a relationship between pure space and the time of movement.

Beyond physics all other special sciences operate with an analogical concept of motion. A physical motion is an expression of energy and implies therefore immediately the causal relationship, something that is not the case with pure motion. Biology operates with the concept of development, psychology with the concept of the motion of feeling or emotion, logic with the concept of logical progression of thought, historiography with the concept of historical development, economics with the concept of economic mobility, etc.

In opposition to the concept pure motion the analogical concepts of motion are always qualified by the nucleus of meaning of an aspect of experiencing within which motion functions as an analogical moment.

I point further to the concept of life. This concept can be employed without further modal qualification only in biology. If one speaks here of organic life, then the adjective "organic" is not a modal nucleus of meaning which qualifies the aspect of life; on the contrary, it is an analogy of the concept of number: the unity in an interconnected multiplicity of life-functions which are first of all qualified by life itself. This analogy is unbreakably connected with a spatial analogy: the whole and its parts, and a physical analogy: the organizing impulse of life or energy of life. In the other sciences the concept of life is always qualified by the nucleus of meaning of the particular aspect of experience in question. Psychology speaks of feeling-life, historiography speaks of cultural life, linguistics speaks of the life of language, economics speaks of economic life, aesthetics speaks of aesthetic life, jurisprudence speaks of the life of law, theology of the life of faith, etc.

Here we encounter at the same time various aspects of human society which it is impossible to bring under a single concept of life, for instance, the biological or the psychological. The life of language, for instance, with its inherent development, expresses itself in the super-sensory social aspect of the understanding of expression and meaning in a syntactical context, in which the syntactical whole conditions the meaning of the parts of speech. Understanding as such is indeed an individual act of consciousness, but the structure of a language and its inherent development can never be explained in terms of the individual consciousness. They have a super-individual, social character, by reason of which the life of language itself is an aspect of human society and understanding as an individual act of consciousness is possible only within the sphere of language of that society and is conditioned by it. Those who wish to reduce terms such as living and dead languages to simple metaphors would have to point out where the secondary usage of the terms life and death begins and the original meaning ends. Furthermore he would have to be in a position not only to replace the term "the life of language" but also to replace such expressions as the life of science, historical life of culture, social life, the life of law, etc., by others which in their significance do not display any real connection with what he understands by life in its primary sense. But such an attempt would be doomed to failure, because then the expression "human society" could not be retained. Because if one eliminated all of the indicated modal aspects as modalities of *life* from the society, then there would be nothing human left to society at all.

Should one say that in its proper sense human life can be ascribed only to the individual person and that human society is nothing else than an interaction between living individuals, then one would not have advanced one step with respect to the problem concerning the understanding of life in the primary sense. In any case, even apart from the analogical and therefore ambiguous character of the concept of interaction, the individual human life itself manifests all of the modalities which are given in the various aspects of our experiencing, and which—in spite of their mutual irreducibility—are interrelated in an unbreakable coherence of meaning. And to this belong also the historical aspect of culture, the aspect of language, the aspect of social intercourse, the economical, the juridical aspect, etc., etc., which are all indispensable aspects of *living-together*, and which belong to the human horizon of experiencing.

The current distinction between bodily and spiritual life does not bring us any farther, because aside from the question whether this distinction has any meaning within the *temporal* life of man, it lies in any case on another niveau than that of the modal structure of the aspects of experience. For this structure is not characterized by concrete things or entities, which participate in the various modal aspects of experience. The structure of a modal aspect is something else than the typical totality-structure of an individual thing.

Still another example of analogical usage is presented by the concept of power. Power in the proper sense of the word or the imposition of form according to a free project is, as I remarked earlier, the nucleus of the historical aspect of culture. But systematic, theoretical analysis discloses within the logical aspect the analogical moment of *logical* control, which displays an inner connection with the concept of power in its original historical sense without being reducible to it. In jurisprudence we work with the concept of juridical power. Juridical power is indeed founded in historical-social power, but it cannot be identified with it without eliminating the juridical itself. Also the concept of economy has an analogical character in so far as it is employed outside of the science of economics. Logical economy of thought, technical economy, lingual economy, aesthetic economy, economy in the forms of social intercourse, juridical economy, are separate analogies of the original economic aspect of our experiencing.

A careful analysis of the analogical concepts, of which I have given here only a few examples, is fruitful and necessary to the highest degree, in the first place because it preserves us from false problematics in philosophy as well as in the special sciences.

In epistemology the lack of analysis of the analogical concepts has constantly led to the attempt to discover in one particular aspect of our horizon of experience the origin of the other aspects, or at least a portion of them. Epistemological investigation began, in other words, with an absolutizing of particular aspects of our experiencing which were thereby wrenched out of the coherence of meaning with the other aspects. In this fashion the way was at once cut off to an insight into the structure of our horizon of experience with its integral coherence of all aspects, and to distinguishing the analogical moments from the original nuclei of meaning in the concepts of the special sciences. In this fashion the entire problem of the analogical concepts was eliminated.

Even Kant's *Critique of Pure Reason* rests upon an uncritical usage of these concepts, whose analogical character he did not recognize.

Kant's theoretical conception of human experience and of empirical reality is exclusively oriented to the physics of Newton and its mathematical foundations. For Kant experience has only two aspects: the sensory and the logical. The transcendental structure of the horizon of experiencing which has been truncated in this fashion is then constituted by him by way of four classes of transcendental-logical categories of thought, in which he distinguishes the mathematical and the dynamical, and two *a priori* forms of sensory perception, namely, space and time. The fundamental concepts of pure mathematics (he names those of number, spatial extension and spatial figure) are now supposed to have risen by way of a schematizing of the mathematical categories of our logical function of thought in the sensory forms of perception. Those of empirical physics, which are supposed to be related *a priori* to a sensory material of experience, namely, that of the real constancy of change, physical causality, and reciprocity in time, are supposed to have originated by a similar schematizing of the dynamical categories. By schematizing is understood here an *a priori* connection or synthesis of the pure concept of understanding with the pure sensory form of space or perhaps time, by means of which we produce on the concept of the understanding a sort of *a priori* image or "monogram." According to Kant this is the product of the transcendental power of imagination; but it proceeds nevertheless from the logical function of thought, because only this is supposed to be capable of binding a multiplicity into a unity.

All of the fundamental concepts of mathematics and mathematical physics according to the founder of the critical philosophy are grounded therefore in the *a priori* structure of our experience. Because he accepted only two aspects of experience: the sensory and the logical, he could not arrive, however, at an analysis of the proper sense of these fundamental concepts. And this indeed has fateful consequences for the critical value of Kant's epistemological investigation.

I want to elucidate this statement by means of various illustrations. The first class of what Kant calls the mathematical categories comprises the logical concepts of unity, multiplicity, and totality. These categories which are themselves strictly logical are schematized in time as pure forms of perception. Out of this the concept of number is supposed to arise, because I successively add new unities of a like nature to the temporally conceived logical unity. But what meaning now does this logical unity, multiplicity, and totality have? And in what sense is the notion of time employed here? Kant names the categories of the first class categories of quantity. And he means here by quantity apparently the *how much*. The first question that rises is this: whether the concept quantity can have an original logical sense. And this question involves the second: as to what sense can be ascribed to the logical as such and whether there are fundamental logical concepts (*stambegrippen*) which are purely logical, that is to say, that are able to be conceived apart from their interconnection with the non-logical aspects of our experiencing.

With regard to the first question Kant, as he

himself explained, derived his categories of quantity out of the logical form of the judgments which contain an expression of general, particular, or individual character. If we consider the three judgments: All men are mortal, some men are mortal, and Socrates is mortal, only as to their logical extension in abstraction from all other content, then according to Kant we have deduced the pure forms of understanding of unity, multiplicity, and totality. As pure *a priori* forms of our logical function of thought they have according to him a purely logical character, and logical is understood by him in the current meaning of *analytical*. As an additional consideration he remarks that they bring under a concept a pure synthesis of a multiplicity which is given *a priori* in the sensory forms of perception of space and time. That is to say that according to their nature they are related to space and time as *a priori* forms of sensory experiencing which themselves are not of a logical character. It is precisely this that gives them in Kant's transcendental logic their epistemological significance. But this is already to admit that logical unity, multiplicity, and totality cannot have any analytical sense without being connected with a multiplicity which apparently is already given in the pure forms of space and time of sensory perception and therefore cannot be logical in character.

This multiplicity must reside, therefore, in the *a priori* structure of space and time itself and must be irreducible to the analytical category of multiplicity. Thereby the word "multiplicity" becomes ambiguous; it assumes an analogical character. It can signify logical multiplicity; but it can also refer to a spatial or temporal multiplicity which underlies the logical and to which the logical is related in an *a priori* fashion. Now the logical concepts of unity, multiplicity, and totality have a unique analogical character because their modal sense is qualified analytically. A logical unity is not as such a quantitative thing,[4] any more than logical multiplicity and totality are. The addition of a logical unity to another logical unity can at most produce a new logical unity but never the number 2.

According to its analytical aspect a concept is a logical unity in an analytical diversity of characteristics. The analytical relationships of identity and diversity,

---

4. In his *Critique of Pure Reason* (transcendental logic, first book, third section, paragraph 12) Kant himself must recognize this, when he writes: "In all knowledge of an object there is unity of concept, which may be entitled *qualitative unity*, so far as we think by it only the unity in the combination of the manifold of our knowledge. . ." (Norman Kemp Smith, tr., *Immanuel Kant's Critique of Pure Reason*, London: Macmillan, 1933, p. 118). And yet Kant calls this same logical unity a *category of quantity* "in its formal meaning!" This is a confusing play with words.

THE ANALOGICAL CONCEPTS 27

implication and exclusion, do not have a quantitative sense. They are not relations of addition and subtraction, multiplication and division, etc. But they presuppose the original experiential aspect of multiplicity.

If I construct the hypothetical logical judgment of relationship: if P then Q and Q implies R, then it holds that P implies R, then I have without doubt introduced three terms. But this number 3 cannot be deduced out of the analytical relation of implication but only out of the quantitative relationship which is a presupposition of each analytical relation and which is ranged under another, irreducible modal aspect of our experience, namely, that of quantity.

The sensory multiplicity in space and time, which Kant viewed as presupposed by logical unity, multiplicity, and totality, also cannot have a purely quantitative but only an analogical character, because it is qualified in a sensory fashion. I cannot add, subtract, multiply, and divide sensory impressions any more than I can do this from a logical point of view with the terms of an analytical relationship. If I ascertain that I have seen four flashes of light from a signal, then I can never claim that I have experienced the number 4 in a sensory fashion. The sensory relationships of similarity and dissimilarity in which we grasp a multiplicity of impressions are never of a quantitative nor of a logical-analytical character. If therefore neither the logical categories of unity, multiplicity, and totality, nor sensory multiplicity have an original quantitative sense, then it is also impossible for number to originate out of a connection of the logical and the sensory functions of experience.

Now Kant says, however, that number originates because we give an *a priori* schema to the logical categories of quantity in time as the sensory form of perception.

But time[5] has as just as many modalities as human experience has aspects. It is only in the aspect of number that it manifests the quantitative sense of an order of arithmetical more and less which comes to expression in the relations of plus and minus. In the sensory aspect of feeling, on the contrary, the order of

---

5. Universal time manifests an unbreakable correlation of temporal *order* and factual *duration*, which are not reducible to each other. Since the latter is bound to the level of concrete event and since both logic and pure mathematics abstract from it, the appearance arises that they operate with a concept of order that is not of a temporal character. In reality they operate with modalities of the *order* of time to which there always correspond in our concrete experience modalities of *duration*. Cf., my article, "The Problem of Time in the Philosophy of the Cosmonomic Idea" ("Het tijdsprobleem in de Wijsbegeerte der Wetsidee," *Philosophia Reformata*, V (1940), 160-182, 193-234).

THE ANALOGICAL CONCEPTS 29

time can never reveal itself in relationships of quantity but only in successive relationships of the intensity of feeling.

Kant, however, speaks of time in general, without taking into account the modal ambiguity of the concept of time. Undoubtedly there is time in the sense of something that embraces all of the modal aspects of our experience, which expresses itself in each of these aspects in a particular modality without exhausting itself in any one of them. We cannot grasp this universal time in a concept because it is what makes every concept possible. We can only grasp it in concrete experiencing (*beleven*). But in the theoretical attitude of thought in science, in which we abstract from the continuous connection of this universal time in order to articulate its modal aspects in an analytical fashion, we encounter time only in the multiplicity of its modal aspects. We can and must form a theoretical concept of the various aspects of time if we do not wish to fall prey to the ambiguity of the term "time" in its theoretical employment.

Parenthetically, there is this difference in principle between the analogical concept of time and the analogical concept of space, that there is no single aspect of experience in which time appears without modal qualification. That is to say, time does not appear as a particular aspect in the horizon of human experiencing, while space does. If we can judge by his definition, Newton's absolute time was nothing else than the kinematic aspect of time conceived as pure movement in a straight line: tempus quod aequaliter fluit. Universal time lies on a level much deeper than the spatial aspect of experience. The spatial aspect is *itself* an aspect of time in which universal time expresses itself in the complete simultaneity of all spatial positions[6] as the basis for pure time of movement. When Kant thus argues that space and time cannot be concepts because there is only one space and one time of which all particular spaces and times are simply parts, then this argument is uncritical because he does not observe the ambiguity of the terms space in time in their theoretical use. Without doubt time and space are not themselves concepts, but their distinct modal meanings must be kept separate by means of concepts.

What now did Kant understand by space and time as pure forms of sensory perception? With the term "space" he had in mind, as we saw earlier, the three dimensional space of Euclidean geometry. With the term "time" he had in mind time in the sense of the pure or mathematical notion of movement, that is to say, what is called the kinematic. Newton had

---

6. This was already noticed by Plato in his dialogue *Parmenides*.

THE ANALOGICAL CONCEPTS 31

called them absolute or mathematical space and time and had assigned to them a metaphysical significance. Kant first transformed them into *a priori* concepts of the understanding and in his truly critical period into *a priori* forms of perception which are the ground of the possibility of all sensory perception.

Now we have already noted that pure Euclidean space can never belong to the same aspect of our experience as the space of sensory perception and that in the former we can never receive sensory impressions of space any more than exact figures of space can appear in the space of sensory perception. And it is just as possible that in the exact kinematic time of movement in a straight line our subjective feelings and sensory observations could not exist because neither of them are pure extensive movements. For this reason the temporal modal aspect of movement cannot be a sensory form of perception any more than pure Euclidean space can be.

If we look, finally, at Kant's category of causality, we see, as he himself maintains, that it is deduced from the hypothetical judgment according to its logical form of *ground and consequent*. Now this logical concept of causality is a distinct analogical concept because it has an analytical qualification and takes on another sense in another modal qualification. How can it be then, as Kant expressly claims, the origin of the concept of power and of the relationship of action and passion? Kant maintains that these concepts, which are deduced exclusively from the category of causality, are pure concepts of the understanding.[7] It is however undeniable that in the analytical relationship of cause and effect the original concept of power or energy cannot be conceived and that it is also impossible for the latter to arise out of an *a priori* synthesis of this logical relationship with time in a kinematic sense, which Kant unjustifiably called a form of perception. The concept of energy can only be related to an original modal aspect of our experience, that of energetic effect, in which the causal relationship appears for the first time and in its original meaning. But this aspect is neither of a sensory nor of a logical character. Here

---

7. *Critique of Pure Reason*, transcendental logic, first book, third section, paragraph 10: "If we have the original and primitive concepts, it is easy to add the derivative and subsidiary, and so to give a complete picture of the family tree of the (concepts of) pure understanding. . . . It can easily be carried out, with the aid of the ontological manuals—for instance, by placing under the category of causality the predicables of force, action, passion. . ." (Tr. Norman Kemp Smith, *Immanuel Kant's Critique of Pure Reason*, pp. 114-115). The terms "Handlung" and "Leiden" which are translated in the text by "action" (*werking*) and "passion" (*ondergang van de werking*) are derived by Kant from the Aristotelian doctrine of the categories.

again Kant's *Critique of Pure Reason* appears to rest on a misapprehension of the analogical character of the categories and the so-called forms of perception which were introduced by him.

We have already remarked that the dogmatic orientation of this criticism,[8] in which the great thinker of Königsberg maintains without any further proof that all human experiential knowledge can arise out of two sources only, namely, sensation and understanding, prevented him from recognizing the analogical structure of the aspects of experience. Because it is only in the integral connection of all these aspects that they can disclose themselves to our analysis. As soon as the sensory aspect of feeling and the logical aspect are isolated in an epistemological fashion from this connection, we lose to sight their modal structure of meaning.

* * * * *

One cannot avoid the ambiguous character of the analogical concepts of the various sciences by subjecting them to a logical formalization with the aid of symbolic logic. The opinion which is maintained

---

8. That this dogmatic tendency is completely dominated by the dialectical religious ground motive of nature and freedom I have shown in Vol. I. of my work *A New Critique of Theoretical Thought* (Philadelphia: Presbyterian and Reformed Publishing Co., 1953).

in the circle of so-called scientific empiricism, that all concepts of science are essentially of one kind, rests upon an epistemological prejudice which cannot give any account of the true state of affairs. Even the analytical relations manifest analogical moments. In the analytical extension of a concept or a judgment there lurk both spatial and numerical analogies. Misled by these analogies mathematical logic in the form given it by Russell and Whitehead has tried to deduce the concept of number in a purely analytical fashion out of the concept of class, by defining it as a class of equivalent classes, while in reality they placed the concept of number at the foundation of the concept of class.

* * * * *

The Philosophical Neglect

Finally I want to answer the question how it can be explained that the fundamental relationship of the analogical concepts to the structure of the horizon of human experience has been so largely ignored by philosophy. This question is even more imperative because already in Greek and scholastic philosophy much attention was given to the analogical concepts.

Here they were correctly distinguished from generic and specific concepts, which indicate the attributes which the things subsumed under them have in common really, that is to say, in an *unambiguous sense*. On the contrary, the analogical concepts have to do with predicates which are applicable to things only in a secondary fashion, that is to say, in line with their different natures and thus in various ways.

The traditional Aristotelian-scholastic logic dealt with analogies in connection with the ambiguity of words, and here the true analogy was sharply distinguished from the metaphor. But the distinction between them was founded in the final analysis metaphysically in the order of being. *Being* as the most fundamental of all concepts, which lies at the foundation of all others, became the analogical concept par excellence and all general, so-called transcendental determinations and distinctions of being participated in this analogical character.

Precisely this metaphysical interpretation which was given to the analogical concepts was bound to detract attention from their relationship to the modal aspects of the horizon of human experience.

It is important at this point to investigate briefly the origin of the doctrine of the *analogia entis*, which in contemporary theology again came to stand in the center of attention as a result of Karl Barth's attack on it. Because by this means a sharp light is thrown on the fact that the philosophical view of analogy stands in immediate connection with the religious starting point of thought.

In the interests of time I must limit myself here to supporting this statement by reference to Greek thought, even though it also holds true for modern philosophy.

As I have tried to show in the first volume of my book *Reformation and Scholasticism in Philosophy*[9] on the foundation of an investigation of the sources, Greek thought was dominated by a central ground motive that since Aristotle has been called the form-matter scheme. It had its origin in an irreconcilable conflict in the religious consciousness of the Greeks between an older religion of life and the younger culture religion of the Olympic pantheon. In the first the deity was not conceived in a personal form. Instead it was seen as an eternally flowing stream of life out of which have arisen periodically the successive generations of beings, who have sought to establish their individual form and characteristics but precisely for this reason have been subjected to a terrible fate, *anangke* or *heimarmene tyche*. In the religion of Dionysus which was imported from Thrace this religion of life found its most significant

---

9. *Reformatie en scholastiek in de wijsbegeerte*, I (Franeker: T. Wever, 1949) (Tr.)

embodiment. This cult finally resulted in the orgies, in which according to the accounts only women took part,[10] who with wild abandon tore apart an animal and ate the flesh raw. These orgies led to ecstasy at the approach of the god Dionysus by means of which the soul as a principle of life was supposed to break through the limits of the body in order to unite itself with a flowing stream of universal life. The rending of the animal's body obtained thereby a symbolical meaning.

One might well express the fundamental idea of this symbolical act with a variation of Mephisto's statement in Goethe's *Faust*:

> *Denn alles was geformt entsteht.*
> *Ist wert das es zu Grunde geht.*

This conception of the ever-flowing stream of life which breaks through every form corresponds to the Greek motive of matter in its original signification. In the fifth book of his *Metaphysics* Aristotle informs us

---

10. See Martin P. Nilsson: *Geschichte der Griechischen Religion*, I (München, 1941), 537: "die bakchischen Orgien erscheinen ausschlieszlich als eine Sache der Frauen." Compare also the still classical description of these orgies by Rohde: *Psyche* II, 1 ff., which is only out of date with regard to its view that the belief in immortality has its origin in the Thracian cult of Dionysus.

that *physis* or nature has been indeed conceived in a one-sided fashion in the sense of the origin of things which undergo development, or of the origin of their movement, by which he undoubtedly referred to the Ionian philosophy of nature. The word *physis* stands in any case in connection with the verb *phuesthai*.

The motive of form on the other hand pertains to the more recent religion of culture of the Olympic gods, the religion of form, measure, and harmony. The Olympic gods leave mother earth with its eternally flowing stream of life and threatening *anangke*. They have a super-sensory personal form and are immortal. But they have no power over *anangke*.

For this reason the Greeks in their private lives held fast to the older religions of nature and of life. The Olympic religion became only the official religion of the city state.

Even after the mythological form of these religions was undermined by philosophical criticism the form-matter-motive, which was born out of the conflict between the older and the more recent religions, still dominated Greek thinking, just as it did Greek art[11]

---

11. Aeschylus' famous trilogy "Orestea," for instance, concentrates on the same problem that Plato treats in his Timaeus, namely, the relationship of blind *anangke* to the rational form-giving power of divine thought. Plato as well as Aeschylus looked for the solution in a "persuasion" of *anangke* by the

THE ANALOGICAL CONCEPTS 39

and Greek society.[12] It had an inner dialectical character and had the tendency to drive in diametrically opposed directions the theoretical thought that had come under its sway.

In the absence of the possibility of finding a true synthesis between the mutually antagonistic motives there remained only the possibility of assigning to one of them the primacy or religious precedence, which

---

divine *nous*. Compare the end of Aeschylus' tragedy, "So Zeus and necessity were reconciled."

12. The Greek city state is the unit which carried the Olympic religion of culture. According to the Greek conception, which was for the first time worked out in a philosophical way by Protagoras, man, who as a natural being is still caught in the untrammeled wildness of the principle of matter (the *rheuston*), first becomes a complete man by the forming power of the city state. Aristotle later expresses this in such a fashion, that anyone who lives outside of the city state must either be a god or an animal.

Upon this rests also the contrast between the Greek and the barbarian. Since the latter, living outside of the city state, does not fully attain to the proper essence of man, he is by nature destined for slavery. The Hellenistic Stoics were the first who broke with this view of man which was oriented to the city state (*polis*). So the Greek conception of *humanitas* and the central role of the *polis* in training for it is completely dominated by the form-motive of the religion of culture in its diametrical opposition to the matter-motive of the older religion of life. Also it is the Delphic Apollo who as the law-giver subjects the ecstatic religion of Dionysus to the limiting principle of form of the city state.

then was accompanied by a principal dedivinization of the opposing religious motif. The attempt was also made, however, to bind together the mutually antithetical religious motives by means of a dialectical logic, without possessing a starting point for a true synthesis. In the latter case there arises the dialectical illusion that one has transcended the antithesis in the religious starting point of thought by means of an embracing theoretical concept. A dialectical illusion indeed! Because a truly theoretical synthesis can only be accomplished in terms of a starting point which transcends both terms of the theoretical antithesis and in which they discover their deeper root-unity.

The Greek form-matter motif excludes however a deeper root-unity of the principle of form and the principle of matter, because what is involved here is a dualism in the central religious sphere out of which theoretical thought itself proceeds.

The principles of form and matter were for Greek thought mutually irreducible principles of the origin of the cosmos and the question which of them was thought to be divine depended entirely on the question as to which of them was assigned the religious primacy. The Ionian philosophy of nature arose in a time of religious crisis in which the old religion of life openly opposed the official religion of nature

THE ANALOGICAL CONCEPTS 41

in the well-known Dionysian movement. And this philosophy of nature developed under the primacy of the principle of matter. In his *Metaphysik des Altertums* Stenzel correctly remarks that these philosophers are dominated by the effort to deprive the world of its form. The true *Arche*, the divine origin of all things, is here indeed the eternally flowing stream of life, for the most part represented by the symbol of a so-called dynamic element. Anaximander called it the *apeiron* and presented the individual existence of things as a guilt that must be atoned for in the order of time: "The origin of all things is the *apeiron*. Into that from which things take their rise they pass away once more, as is ordained, for they make reparation and satisfaction to one another for their injustice according to the ordering of time."

But already in the first phase of Greek thought the dialectic of the motif of form and matter leads to a polar antithesis. It is Parmenides, the founder of the Eleatic school, who founds the metaphysical doctrine of being and conceives this *Being* in the divine spherical form of the firmament. No true being can be ascribed to the material principle of eternal flux. As non-being becoming cannot be thought in a logical fashion. The attributes of truth and unity were attributed to Being.

That here indeed the religious motif of form sets

itself over against the material principle of eternal flux appears from the ceremonious fashion in which Parmenides declares that *anangke* and *dike* hold being within the established limits of its spherical form and guard it against every transgression of these boundaries, by which it would plunge itself into the deceptive temporal stream of becoming. But the motif of form does not come to expression here in the pure sense that it does in cultural religion. Likely under Orphic influence it finds itself bound up with the old ouranic motif of the reverencing of heaven.

But already in Anaxagoras this form-motif has freed itself from these ouranic influences and is conceived again in the pure sense of cultural religion. Culture is the controlling manner of giving form to a material according to a free project, and in this it distinguishes itself in principle from the kind of forming that we encounter in nature. Now Anaxagoras is the first who seeks the divine origin of all cosmic form in the *nous*, the divine power of thought which is completely unmixed with matter because otherwise it would not be able to control (*kratein*) it. Matter is completely deprived of all divine character. Even the principle of the eternally flowing movement of life is denied to it. It becomes an inert chaos, in which the basic elements of all things are mixed together. From the divine spirit proceeds the first form-giving movement which makes out of chaos a cosmos. So the divinity is conceived as the demiurge, the divine builder, who does not create but simply gives form to an available material which has its own independent existence.

This conception which in Anaxagoras was not worked out consistently in his philosophy of nature, was taken over by Socrates, likely by way of Diogenes of Apollonia, and in his thinking obtains an ethical, aesthetical turn. The divine *nous* has given form to everything in the visible cosmos according to the idea of the *kalokagathon*. Everything that has form is an expression of this idea of the good and the beautiful. It is for a good purpose, it has an *arete*, it corresponds to a teleological order which is oriented to the realization of the good and the beautiful.

In the first conception of Plato's doctrine of the ideas, as we find it in the dialogue Phaedo, the *eide* or ideal forms of being are yet conceived strongly in an Eleatic sense. They are altogether simple, underived and indestructible, completely immobile, and nevertheless they are represented as the true foundations of being (*aitiai*) of the perishable things of the sensory world. The latter participate (*methexis*) in the eternal forms of being which are their *paradeigmata*.

But the sharp distinction which is adopted here

between the transcendent world of eternal forms and the world of sensibly experienceable things which is subject to the principle of matter made the problem of participation (*methexis*) in this form insoluble.

If according to the model of the Eleatic form of being the *eidos* is an absolute unity without multiplicity and can have no relationship at all to the principle of matter of the eternally flowing movement of life, how can it then be the cause of the perishable world of form and how can it as an absolute unity become a diversity in the sensory world?

Here is the problem that holds the attention of Plato in his so-called "Eleatic Dialogues," *Parmenides*, *The Sophist*, and *The Republic*. In principle it is the problem how theoretical thought can discover a synthesis between the principle of form and that of matter. In order to get a true synthesis one would have to be able to refer both of these principles back to a higher root-unity.

The material principle of the eternally flowing stream of life was in essence oriented to the organic aspect of life of the world of experience. The principle of form of the religion of culture was oriented to the historical aspect of culture. In our temporal horizon of experience both aspects are unbreakably interconnected and are related to the I as the root-unity of all temporal aspects. Without the movement of life no cultural giving of form is possible and within the historical aspect of culture this connection comes to expression in the historical development of the life of culture, which is a true analogy of the organic development of life. But in the ancient religion of life the organic aspect of life was absolutized and divinized, just as in the later religion of culture the historical aspect of culture was absolutized, so that it lost its inner relationship to the organic aspect of life. It was just for this reason that the form-matter motif, which arose from the encounter between these two religions, did not allow for any true synthesis between the two antagonistic motives.

In spite of this fact Plato attempts in his Eleatic Dialogues to discover such a synthesis with the aid of a dialectical logic. To this end he introduces a number of dialectical ideas which have the purpose of bringing into logical correlation the principles of form and matter which in themselves are mutually irreconcilable. The highest and most all inclusive dialectical idea now becomes that of *Being*.

Parmenides attained to his metaphysical concept of being by way of absolutizing the logical relation of identity (*estin einai*) and the concept of logical unity. According to him, logical *thought* and logical *being* are always identical. In his dialectical concept of being Plato proceeds from the correlation between logical identity and diversity and between logical unity and diversity. Every definition of what a thing *is* implies an infinite series of predicates which are *not* attributable in a logical sense to the thing.

Being and *non-being* are thus logical correlatives. Although the flowing stream of becoming is called by Parmenides a *non-being* that is excluded by *Being*, the dialectical-logical idea of being can also embrace this non-being. It connects both the "*on*" and the "*me on*," rest and movement, identity and difference, unity and diversity.

Here we have the origin of the famous analogical concept of being. In the dialogue *Philebus* the stream of becoming is conceived of as a *genesis eis ousian*, a process of becoming a being. By this means the later Aristotelian distinction of potentiality and actuality was anticipated in principle and at the same time there was introduced into the doctrine of being the teleological orientation which attributes to matter a striving towards form as the natural good of the composite essence.

So to the Eleatic qualifications of being were added unity and truth, which were the Socratic qualifications of the good. But all of these qualifications now obtained an analogical character in order to bring about a dialectical synthesis between the two antagonistic principles of form and matter. I have pointed out already that this synthesis could only be an illusory one.

The analogical doctrine of being, as it was worked out by Aristotle, has two poles: God as pure actual form and primal matter as the principle of incompleteness. In the Greek doctrine of being, which did not know the idea of creation, it was never possible to reduce the principle of matter to the principle of form. The Aristotelian god is no creator but only the ideal goal of all purposeful activity in the cosmos.

So the analogical concept of being took its rise under the influence of a dialectical religious ground-motif, which also impresses its dualistic stamp on the entire theoretical view of temporal reality. Even in the Aristotelian doctrine of the categories, which was supposed to give to the analogical doctrine of being is more precise formulation, the dualism of the form principle and the material principle also comes to expression. The categories of quantity, including those of spatiality and size, are limited exclusively to matter, and the logical activity of thought, the *nous poietikos*, is made completely independent of the perishable material body. The dichotomy which was thereby introduced into the human horizon of experience made it impossible to gain an insight into the integral coherence of meaning of all modal aspects of our temporal experience and therefore also into the integral significance of analogy.

The analogies which manifest themselves in all of these aspects presuppose a deeper root-unity in the central religious sphere of our existence in which all modal differentiation of meaning must be directed concentrically to the absolute Origin of all meaning.

The Greek doctrine of *analogia entis* cannot refer the analogies to a religious root-unity because its religious starting point excludes this radical unity. Here analogy has the last word. This means that this analogical concept of being in its intrinsic ambiguity cannot provide any useful foundation for insight into the mutual relationship and inner connection of the various aspects of experience which is expressed in their analogical structure. So instead of a *primum notum* it becomes an *asylum ignorantiae*.

Within the temporal horizon of human experience analogy is the expression of an unbreakable coherence of meaning in an irreducible diversity of meaning. Both presuppose a deeper unity of meaning in the religious center of human existence. And that unity of meaning in its turn is simply the creaturely expression of the divine unity of the Origin, which transcends all diversity of meaning and thus every analogy and which

THE ANALOGICAL CONCEPTS 49

exclusively is Being. There is no analogical concept of being that would be able to embrace both the creature and his divine Origin. Analogy is exclusively of a creaturely nature.

ABOUT THE AUTHOR

Herman Dooyeweerd (1894-1977) was born in Amsterdam to Calvinistic parents whose convictions and way of life were profoundly influenced by Abraham Kuyper, the great Dutch statesman, educator and journalist, and one of the protestant leaders through which the evangelical wing of Dutch reformed protestantism emerged. Dooyeweerd is recorded to have had a prolific career as a researcher in philosophy, during which he wrote various profound literary works such as *The New Critique of Theoretical Thought*, *Roots of Western Culture*, *In the Twilight of Western Thought*, and more. He is, without a doubt, one of the most important philosophers that the Netherlands has ever produced, comparable only perhaps with Baruch de Spinoza.

PAIDEIA MONOGRAPHS

Other Titles (2020–):

The Development of Calvinism in North America
H. Evan Runner

Point Counter Point
H. Evan Runner

The Radical Christian Facing Today's Political Malaise
H. Evan Runner

Christ and Christianity
Herman Bavinck

The Secularization of Science
Herman Dooyeweerd

The Concept of Sovereignty in Modern Jurisprudence and Political Science
Herman Dooyeweerd

The Criteria of Progressive and Reactionary Tendencies in History
Herman Dooyeweerd

**Looking for more?
Visit www.paideiapress.ca**

www.ingramcontent.com/pod-product-compliance
Lightning Source LLC
Chambersburg PA
CBHW051959290426
44110CB00015B/2302